SS7 BASICS

by Toni Beninger

Telephony

Division of Intertec Publishing Corp.
55 E. Jackson Blvd.
Chicago, IL 60604

PREFACE

Enhanced services and telephone network features once viewed as luxuries are quickly becoming necessities in today's fast-paced world. And, that's not just true in the business world. Residential telephone service customers as well increasingly are demanding the efficiency and convenience advanced telecommunications technology can provide.

Naturally, telephone operating companies want to satisfy these demands efficiently. Traditional signaling systems cannot provide the level of sophistication required to deliver much more than basic telephone service. To deliver the enhanced services, telecommunications carriers need new, intelligent message-based signaling systems. The solution is Signaling System 7 (SS7). The SS7 specifications are defined in the CCITT Blue Book (1988), Recommendations Q.700-Q.795.

This book is intended to serve as a brief introduction to SS7, allowing the reader to gain an overall appreciation and understanding of its structure and impact. It covers the reasons why SS7 exists and is necessary, as well as step-by-step procedures that describe the actions that occur in the network when SS7 functionality is being used. It is all presented in a straightforward, logical, and easy-to-understand manner, for the technical and the non-technical reader alike.

Special thanks to Bob Duggan of BC Systems Corp. who provided technical insight and valuable commentary, and to Rob McMurtry of Sanctuary Woods who translated ideas and concepts into clear illustrations.

Toni Beninger
Sanctuary Woods Inc.
Kanata, Ont., Canada
July 1991

TABLE OF CONTENTS

1 SS7 DEFINED

Common Channel Signaling is a signaling method in which a signaling channel conveys, by means of labeled messages, signaling information relating to call setup, control, network management, and network maintenance. Examples of Common Channel Signaling systems are CCITT Signaling System No. 7 and various national versions such as the Bell Communications Research (BELLCORE) and AT&T SS7 standards.

WHY SS7?

Worldwide telephone networks are undergoing significant changes as methods of call processing and network management are altered to provide new services and to streamline operations. These changes are driven by user demand for enhanced services and the corresponding efforts of telephone operating companies to satisfy current and future needs. Enhanced services require bidirectional signaling capabilities, flexibility of call setup, and remote database access.

Earlier signaling systems lacked the sophistication required to deliver much more than POTS (Plain Old Telephone Service). These traditional systems use dial pulses and multi-frequency (MF) tones to transmit call and circuit-related information such as dialed digits and circuit busy/idle states.

The complexity of adding new functionality to traditional signaling systems meant that a new network signaling architecture was needed. SS7 was developed to satisfy the telephone operating companies' requirements for an improvement to existing signaling systems.

SS7 NETWORK ARCHITECTURE

A telecommunications network consists of a number of switches and application processors interconnected by transmission circuits. The SS7 network exists within, the telecommunications network and controls it. SS7 achieves this control by creating and transferring call processing, network management, and maintenance messages to the network's various components.

An SS7 network has three distinct components: Service Switching Points, Signal Transfer Points, and Service Control Points. These components may be generically referred to as "nodes" or "signaling points" and are illustrated in Figure 1.1 - SS7 Signaling Network Architecture.

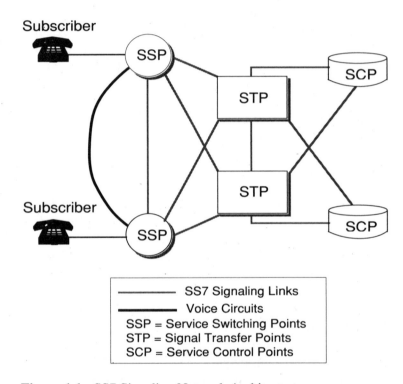

Figure 1.1 - SS7 Signaling Network Architecture

Service Switching Points

Service Switching Points (SSP) are telephone switches interconnected by SS7 links. The SSPs perform call processing on calls that originate, tandem, or terminate at that site. As part of this call processing, the SSP may generate SS7 messages to transfer call-related information to other SSPs, or to send a query to a Service Control Point for instructions on how to route a call.

Signaling Transfer Points

Signaling Transfer Points (STP) are switches that relay messages between network switches and databases. Their main function is to route SS7 messages to the correct outgoing signaling link, based on information contained in the SS7 message address fields.

Service Control Points

Service Control Points (SCP) contain centralized network databases for providing enhanced services. The SCP accepts queries from an SSP and returns the requested information to the originator of the query. For example, enhanced 800 service uses an SCP database to determine the routing on 800 calls. When an 800 call is initiated by a user, the originating SSP sends a query to an 800 database requesting information on how to route the call. The SCP returns the routing information to the SSP originating the query and the call proceeds.

SS7 Reliability

To meet the stringent reliability requirements of public telecommunications networks, a number of safeguards are built into the SS7 protocol:

- STPs and SCPs are normally provisioned in mated pairs. On the failure of individual components, this duplication allows signaling traffic to be automatically diverted to an alternate resource, minimizing the impact on service.

- Signaling links are provisioned with some level of redundancy. Signaling traffic is automatically diverted to alternate links in the case of link failures.

- The SS7 protocol has built-in error recovery mechanisms to ensure reliable transfer of signaling messages in the event of a network failure.

ISDN Access Protocol

SS7 is designed to provide an internationally standardized, general-purpose signaling system; however, SS7 was not intended to be used as the signaling standard for access to the telephone network from PBXs or from telephone sets. To satisfy this latter need, the ISDN-AP (Integrated Services Digital Network - Access Protocol) has been developed. Together, SS7 and the ISDN-AP provide the end-to-end signaling required to deliver enhanced features to users. As an interim step, many telephone exchange carriers use proprietary access signaling to provide enhanced services. Figure 1.2 illustrates the ISDN-AP/SS7 Interface.

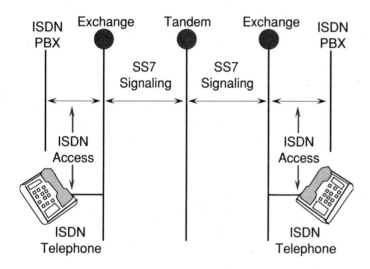

Figure 1.2 - ISDN-AP/SS7 Interface

SS7 MAPPED ONTO THE OSI MODEL

The SS7 standard was designed to map onto the OSI 7-Layer Reference Model, as illustrated in Figure 1.3 - Similarities of the SS7 Protocol and the OSI Model.

The bottom half of the SS7 protocol consists of the Message Transfer Part (MTP). There are three levels to the MTP: Level 1 corresponds to the OSI Layer 1 (Physical Layer); Level 2 corresponds to OSI Layer 2 (Data Link Layer); and, Level 3 corresponds to the bottom of OSI Layer 3 (Network Layer).

The upper half of the SS7 protocol consists of several parts. The SS7 Signaling Connection Control Part (SCCP) corresponds to the top of OSI Layer 3. The ISDN-User Part (ISDN-UP) maps onto OSI Layer 3 as well, and, in addition, it maps onto Layer 4 (Transport Layer), Layer 5 (Session Layer), Layer 6 (Presentation Layer), and Layer 7 (Application Layer). The Transaction Capabilities Application Part (TCAP), the Application Service Elements (ASE), and the Operations, Maintenance and Administration Part (OMAP) of the SS7 protocol all map onto OSI Layer 7 as well.

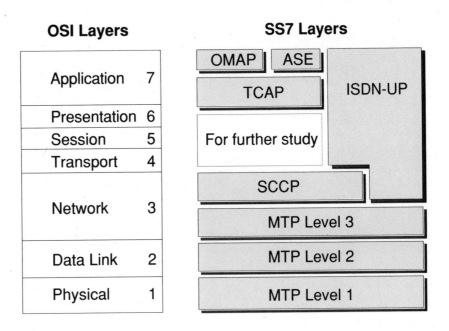

Figure 1.3 - Similarities of the SS7 Protocol and the OSI Model

2 MESSAGE TRANSFER PART LEVEL 1

The Message Transfer Part (MTP) Level is known as the signaling data link. It is equivalent to the OSI model Physical Layer (Layer 1), as illustrated in Figure 2.1 - MTP Level 1 in the SS7 Layers. This layer defines the physical, electrical, and functional characteristics of the signaling links connecting SS7 components.

A signaling data link is a bidirectional transmission path for signaling, comprising two data channels operating together in opposite directions at the same data rate. SS7 signaling data links are capable of operating over terrestrial and satellite transmission links. The signaling terminal at each end of the signaling data link contains the MTP Level 2 functionality for transmitting and receiving SS7 messages.

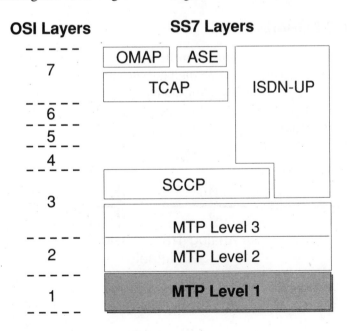

Figure 2.1 - MTP Level 1 in the SS7 Layers

DIGITAL SIGNALING LINK

A digital signaling link is composed of a digital transmission channel connecting two digital switches which provide an interface to the signaling terminals, as illustrated in Figure 2.2 - Digital Signaling Link. The standard rate on a digital transmission channel is 56 kilobits/sec (Kb/s) or 64 Kb/s, although the minimum signaling rate for call control applications is 4.8 Kb/s. Network management applications may use bit rates lower than 4.8 Kb/s.

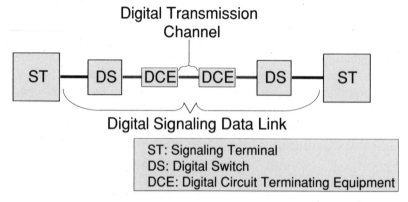

Figure 2.2 - Digital Signaling Link

ANALOG SIGNALING LINK

An analog signaling data link is made up of a voice-frequency analog transmission channel connecting digital switches which provide an interface to the signaling terminals, as illustrated in Figure 2.3 - Analog Signaling Link. While digital signaling links typically use a 56 Kb/s or 64 Kb/s rate, an analog link may use a lower one.

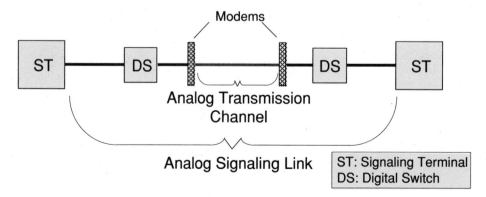

Figure 2.3 - Analog Signaling Link

3 MESSAGE TRANSFER PART LEVEL 2

The Message Transfer Part (MTP) Level 2, together with MTP Level 1, provides a signaling link for reliable transfer of signaling messages between two directly connected signaling points. The MTP Level 2 maps onto Layer 2 of the OSI 7-Layer Model as illustrated in Figure 3.1 - MTP Level 2 of the SS7 Layers.

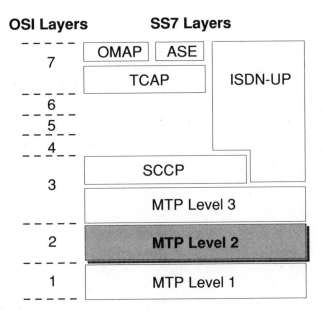

Figure 3.1 - MTP Level 2 of the SS7 Layers

MESSAGE FORMATS

There are three types of signal unit (SU), differentiated by the value contained in the length indicator (LI) field. The signal units are made up of several fields which are defined here to facilitate the descriptions of the actual signal unit types.

F (flag): The beginning and end of a signal unit are indicated by a unique 8-bit pattern, called the flag, which does not appear elsewhere in the signal unit. Measures are taken to ensure that the flag pattern is not imitated elsewhere in the SU. The pattern is 01111110.

CK (cyclic redundancy check): The CK is a 16-bit checksum transmitted with each signal unit. If the checksum does not match at the receiving signaling point, the SU is considered to have errors and is discarded.

SIF (signaling information field): This field contains the routing and signaling information of the message.

SIO (service information octet): This octet is made up of the service indicator and the sub-service field. The service indicator is used to associate the signaling message with a particular MTP user at a signaling point, for instance, the layers above the MTP level. The sub-service field contains the network indicator which is used to differentiate between national and international calls, or between different routing schemes within a single network.

BSN (backward sequence number): The BSN field is used to acknowledge message signal units which have been received from the remote end of the signaling link. The BSN is the sequence number of the signal unit being acknowledged.

BIB (backward indicator bit): The BIB is used in error recovery.

FSN (forward sequence number): The FSN is the sequence number of the signal unit in which it is being carried.

FIB (forward indicator bit): The FIB is used in error recovery.

LI (length indicator): The LI field indicates the number of octets which follow the LI field and precede the CK field.

LI = 0   Fill-in signal unit
LI = 1 or 2 Link status signal unit
2 < LI < 63 Message signal unit

Message Signal Units

The message signal units (MSU), illustrated in Figure 3.2 - Message Signal Unit, carry signaling information for call control, network management, and maintenance in the signaling information field. For example, messages from the Signaling Connection Control Part (SCCP), the ISDN-User Part (ISDN-UP), and the Operations, Administration and Maintenance Part (OMAP) are transferred over the signaling link in the signaling information field of variable length MSUs.

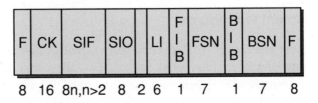

Figure 3.2 - Message Signal Unit

Link Status Signal Unit

The Link Status Signal Unit (LSSU) messages, illustrated in Figure 3.3 - Link Status Signal Unit, provide link status indications to the remote end of the signaling link. Some examples of status indications are: normal, out of alignment, out of service, emergency status.

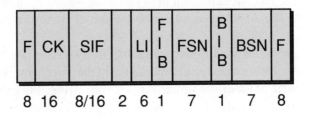

Figure 3.3 - Link Status Signal Unit

Fill-in Signal Unit

The Fill-in Signal Unit (FISU), illustrated in Figure 3.4 - Fill-in Signal Unit, is normally transmitted when no MSUs or LSSUs are being transmitted, allowing the SS7 network to receive immediate notification of signaling link failure.

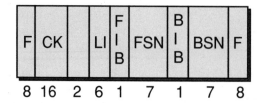

Figure 3.4 - Fill-in Signal Unit

PROCEDURES

Initial Alignment Procedures

Initial alignment procedures (IAP) occur when a signaling link is activated for the first time or restored after a link failure. Two alignment procedures are provided, normal and emergency. A "proving state" is included in both procedures to measure error rates, ensuring that a reliable link is established.

Signaling Link Error Monitoring

Two different types of signaling link error rate monitors are provided to estimate signaling link error rates: the signaling unit error rate monitor (SUERM), and the alignment error rate monitor (AERM).

The SUERM is employed while signaling links are in service. The SUERM provides a fault indication to MTP Level 3 when error thresholds are exceeded. The AERM is used during the initial alignment procedure proving state.

Error rate thresholds are different for each of the two error rate monitors. The error thresholds are based on a combination of the percent message error rate and the length of time over which errors are occurring. For example, at 100% error rate, the threshold would be reached in 128 ms, whereas with a lower error rate, the threshold would not be reached until a longer time had passed.

Flow Control

Flow control procedures are initiated on a signaling link when congestion is detected at either end of the signaling link. Congested conditions may be due to processor outage or link failure anywhere on the network.

The end of the signaling link initiating flow control withholds both positive and negative acknowledgements, and sends a busy status indication to the remote end of the signaling link. If congestion persists, the remote end removes the signaling link from service and initiates an emergency rerouting procedure.

Error Correction

Two forms of error correction are provided, the Basic Method and the Preventive Cyclic Retransmission Method. Both methods are designed to eliminate the possibility of missed, duplicated, or out-of-sequence messages. The Preventive Cyclic Retransmission method is used on signaling links such as satellite links which have large propagation delays.

Basic Method. The basic method of error correction is accomplished by retransmitting those MSUs that were not correctly received in order by the destination signaling point.

Normally, the destination signaling point will reply to a transmitted MSU with a positive acknowledgement. Reception of a positive acknowledgement at the originating signaling point confirms the successful transmission of that MSU. If the exchange sending back the acknowledgement has messages to be sent back at the same time, the acknolwedgement can be contained within the MSU carrying the message.

However, if a negative acknowledgement is returned from the destination signaling point, the originating signaling point will retransmit the MSU and all subsequent MSUs.

Figure 3.5 - Basic Error Correction illustrates the steps in basic error corrections. The steps are described below.

1. Exchange A transmits an MSU with Forward Sequence Number (FSN) = 4.

2. Exchange B acknowledges the successful receipt of the MSU from Step 1 by setting the Backward Sequence Number (BSN) = 4 in the FISU it sends to exchange A.

3 and 4. Exchange A has two MSUs to transmit. FSN = 5 and FSN = 6 are selected and transmitted in order. In this example, the MSU with FSN = 5 was corrupted by transmission errors. Exchange B receives the MSU with FSN=6 correctly.

5.   Exchange B sends a negative acknowledgement to exchange A indicating that the MSU with FSN = 4 was the last MSU successfully received in order. The negative acknowledgement is indicated by toggling the value of the Backward Indication Bit.

6 and 7. Exchange A now retransmits the MSUs with FSN = 5 and FSN = 6, which are received successfully by exchange B.

8.   Exchange B now acknowledges these MSUs by replying with an FISU with BSN = 6. The FISU that serves as the acknowledgement for MSU with FSN = 6 also serves as an acknowledgement for all previous unacknowledged MSUs as well (in this case, MSU with FSN = 5). An exchange can send up to 127 MSUs before requiring an acknowledgement from the remote end.

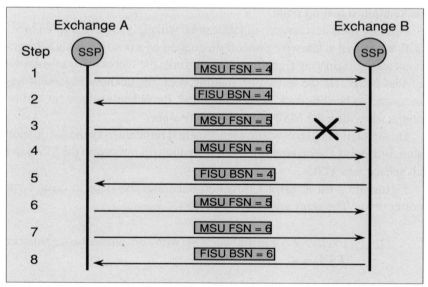

Figure 3.5 - Basic Error Correction

Preventive Method. The preventive method of error correction is accomplished by the local signaling point cyclically retransmitting all the MSUs sent but not yet acknowledged by the remote signaling point.

When there are no new MSUs or LSSUs to be transmitted, all the MSUs that have not been positively acknowledged are retransmitted cyclically.

Figure 3.6 - Preventive Error Correction Method illustrates the steps in the preventive method of error correction. The steps are described below.

1.   Exchange A transmits an MSU with FSN = 4.

2.   Exchange B acknowledges the successful receipt of the MSU from Step 1 by returning an FISU with BSN = 4 to exchange A.

3 and 4. Exchange A sends two additional MSUs to exchange B.

5 and 6. Exchange A has no additional MSUs to transmit and has not received acknowledgement from the MSUs sent in Steps 3 and 4. Exchange A then retransmits the MSUs with FSN = 5 and FSN = 6.

7.   Exchange B acknowledges the MSU with FSN = 6, confirming the receipt of MSU with FSN = 5 as well.

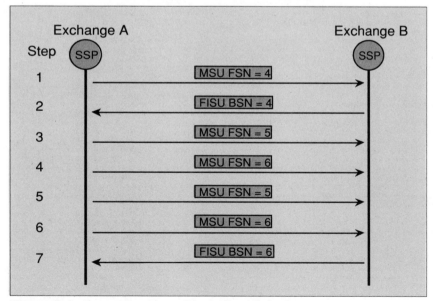

Figure 3.6 - Preventive Error Correction Method

■4 MESSAGE TRANSFER PART LEVEL 3

INTRODUCTION

The Message Transfer Part (MTP) Level 3 provides the functions and procedures related to message routing and network management. MTP Level 3 handles these functions, assuming that signaling points are connected with signaling links as described in MTP Level 1 and Level 2. The MTP Level 3 maps onto Layer 3 of the OSI 7-Layer Model as illustrated in Figure 4.1 - MTP Level 3 of the SS7 Layers.

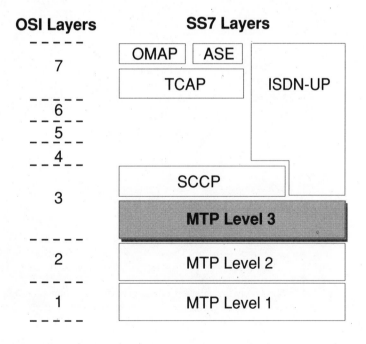

Figure 4.1 - MTP Level 3 of the SS7 Layers

The MTP Level 3 functions can be divided into two basic categories: signaling message handling functions, and signaling network management functions, as illustrated in Figure 4.2 - Signaling Network Functions.

Signaling message handling functions are made up of message routing, discrimination and distribution; these functions are performed at each signaling point in the signaling network.

The signaling network management functions provide the actions and procedures required to activate and maintain signaling service, and to restore normal signaling conditions in the event of disruption in the signaling network, either in signaling links or at signaling points.

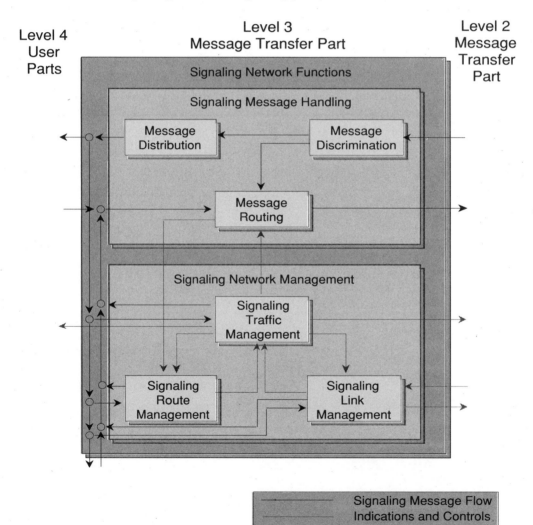

Figure 4.2 - Signaling Network Functions

SIGNALING MESSAGE HANDLING

Signaling message handling is based on the routing label contained in the signaling information field (SIF) of the message signal units. Link status and fill-in signal units travel between two signaling points, therefore neither contains routing labels. See Figure 4.3 - Routing Label Contained Within SIF of the MSU. (The elements of the MSU in this illustration are defined in Chapter Three.) In some cases, the service information octet (SIO) is used for routing purposes as well.

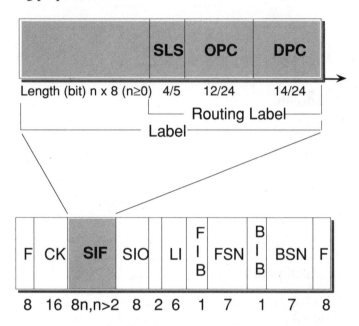

Figure 4.3 - Routing Label Contained Within SIF of the MSU

Signaling Information Field: Routing Label

The standard routing label assumes that each signaling point in a signaling network is allocated a code according to a labeling code plan that is unambiguous in its domain. The routing label includes:

- the originating and destination point codes; and,
- the signaling link selection code.

The originating point code (OPC) indicates the originating point of the message, while the destination point code (DPC) identifies the destination of the message.

The signaling link selection (SLS) field is used for load sharing when two or more links connect adjacent signaling points. Each signaling link is assigned an SLS value. Messages are routed over that signaling link when the MTP 3 sets the SLS field value equal to that of the signaling link. See Figure 4.4 - Signaling Link Selection.

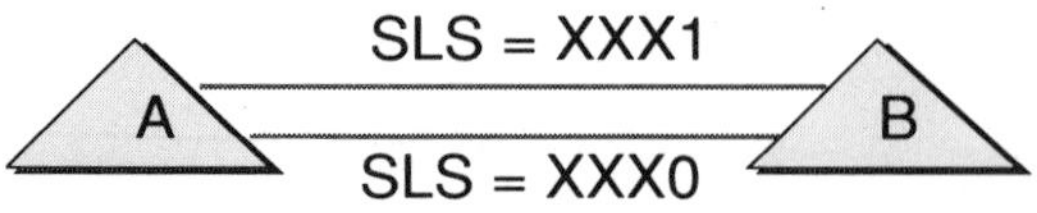

Figure 4.4 - Signaling Link Selection

Message Routing

To route messages, each signaling point has routing information which allows it to determine the signaling link over which a message has to be sent based on information contained in DPC and SLS fields.

Typically, more than one signaling link may be used to carry messages to a given DPC; the selection of the particular signaling link is made by means of the SLS field, thus effecting load sharing.

Message Discrimination

The DPC field of the received message is examined by the discrimination function. The DPC is illustrated in Figure 4.5 - Destination Point Code. If the DPC of the message identifies the receiving signaling point, the message is delivered to the message distribution functions. If the DPC is not that of the receiving signaling point, and if the receiving signaling point has the transfer capability, the message is directed to the message routing function.

Message Distribution

If the message distribution function receives a message from the message discrimination function, then the service indicator is examined and the message is delivered to the appropriate user part.

The service indicator is contained within the SIO field contained in the MSUs, as illustrated in Figure 3.2 - Message Signal Unit, in Chapter Three.

Examples of service indicators are the SCCP, ISDN User Part, Signaling Network Management, and Signaling Network Testing, as illustrated in Figure 4.6 - Message Distribution.

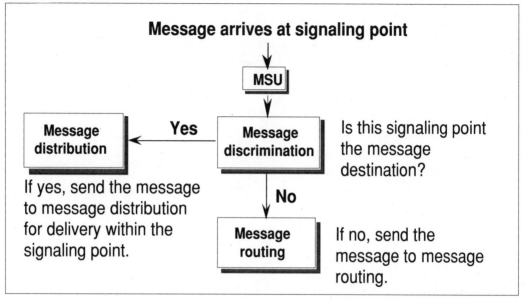

Figure 4.5 - Destination Point Code

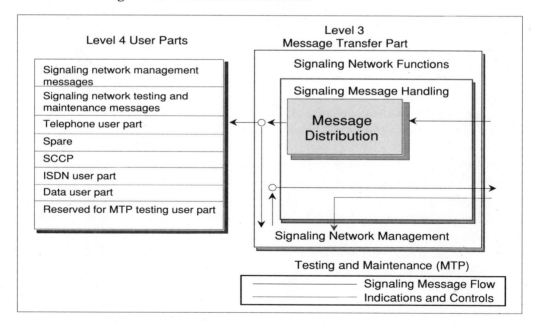

Figure 4.6 - Message Distribution

SIGNALING NETWORK MANAGEMENT

The purpose of the signaling network management functions is to activate new signaling links, to maintain signaling service, to control traffic in case of congestion, and to provide reconfiguration of the signaling network in the case of failures.

In the case of failures, traffic will be rerouted around the failed component if possible, and new signaling links may be activated. Congestion generally results in the change in status of the affected signaling links and routes from an "available" state to an "unavailable" state.

The signaling network management functions are divided into three categories:

- signaling link management;
- signaling route management; and,
- signaling traffic management.

Signaling Link Management

The signaling link management function manages locally attached signaling links. The function is responsible for maintaining predetermined link set capabilities by establishing link sets and initiating action to activate additional links in the event of signaling link failures.

Basic signaling link management capabilities are defined for link activation, deactivation, restoration, and link set activation. Additional capability sets are provided for automatic activation of backup signaling links.

The signaling link management function is carried out by the following procedures: link set activation, signaling link deactivation, signaling link activation, and signaling link restoration.

Link Set Activation: Normal or emergency link set activation procedures are used to establish link sets having no active signaling links.

Normal activation is initiated when a link set is being activated for the first time or when a link set is being restarted and the situation is deemed not to be an emergency. The MTP Level 2 normal Initial Alignment Procedures (IAP) are used on each signaling link in parallel.

Emergency activation is initiated when an immediate re-establishment of the link set is required. The MTP Level 2 emergency IAPs are used.

Signaling Link Deactivation: Active signaling links may be taken out of service for two reasons:

1. The quantity of active signaling links exceeds the predetermined quantity for that link set.

2. Maintenance activities may require a link to be manually deactivated for testing.

Signaling Link Activation: This procedure is used to activate signaling links which are being activated for the first time, or which were previously taken out of service.

Signaling Link Restoration: On detection of a signaling link failure, the signaling link restoration procedure is initiated. This in turn initiates the MTP Level 2 IAP. On successful completion of the IAP, the link is brought into service.

If the IAP fails, a new IAP is repeated until the link is restored, or manual intervention takes place.

Signaling Route Management

The signaling route management functions are used to exchange signaling route availability information between signaling points. Information exchange is accomplished using transfer prohibited, transfer restricted, transfer allowed signaling route set test, transfer controlled, and signaling route set congestion test procedures.

Transfer Prohibited: The transfer prohibited procedure is performed at a signaling point acting as an STP for messages relating to a given destination when it has to notify one or more adjacent signaling points that they must no longer route messages for the given destination via that STP.

The following example lists the steps carried out, and they are illustrated in Figure 4.7 - Transfer Prohibited.

1.　On the failure of links D-F and D-E, D no longer has access to F. D indicates this condition to B and C by sending them transfer prohibited messages identifying that F can no longer be reached via D.

2.　B and C will then initiate the forced rerouting procedure to reroute signaling traffic destined for F via E.

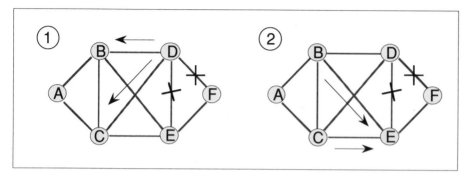

Figure 4.7 - Transfer Prohibited

Transfer Restricted: The transfer restricted procedure is performed at an STP when it has to notify one or more adjacent signaling points that they should not, if possible, route traffic toward a given destination through that signaling point.

Transfer restricted messages are sent under certain link failure and congestion situations. The following example lists the steps carried out, and they are illustrated in Figure 4.8 - Transfer Restricted.

1.　On a failure of link A-B, a changeover procedure is initiated between A and B. Signaling traffic between A and F may take the following path: A-C-B-D-F (highlighted in the illustration). Signaling traffic management may decide that alternate routing over path A-C-D-F is more efficient.

2.　If so, B will send a transfer restricted message to C indicating that C should, if possible, reroute traffic destined for F. In this case, C would initiate the controlled rerouting procedure and route traffic destined for F via D.

3.　Similarly, B will send a second transfer restricted message to D indicating that D should, if possible reroute traffic destined for A.

In this case, D would initiate the controlled rerouting procedure and route traffic destined for A via C.

4. The new route for messages going between A and F is highlighted in Figure 4.8 - Transfer Restricted.

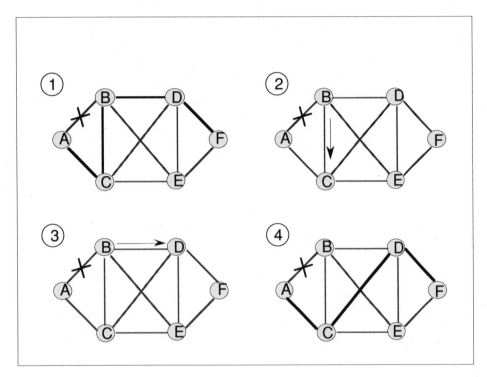

Figure 4.8 - Transfer Restricted

Transfer Allowed: The transfer allowed procedure is performed at an STP when it has to notify one or more adjacent signaling points that they may route traffic toward a given destination through that signaling point.

The following example lists the steps carried out, and they are illustrated in Figure 4.9 - Transfer Allowed.

1. A previous failure of links D-E and D-F would have resulted in B and C rerouting traffic for F via E.

2. On recovery of link D-F, D will send a transfer allowed message to B and C identifying that F can now be reached via D.

3. If D is the primary route for sending traffic from B and C to F, then B and C may then initiate the controlled rerouting procedure to reroute the signaling traffic destined for F via D.

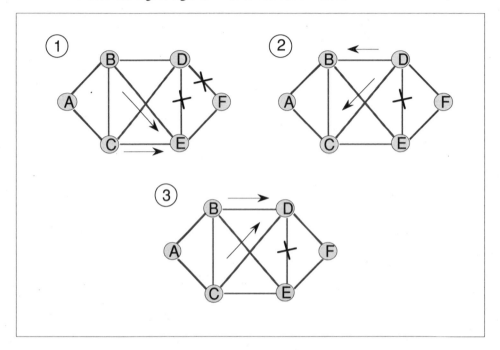

Figure 4.9 -Transfer Allowed

Signaling Route Set Test: This test is performed at signaling points to query whether signaling traffic towards a certain destination may be routed via an adjacent STP. This procedure is activated following receipt of a transfer prohibited or transfer restricted message from an adjacent signaling point.

The following example lists the steps carried out, and they are illustrated in Figure 4.10 - Signaling Route Set Test.

1. A previous failure of links D-E and D-F would have resulted in B and C rerouting traffic for F via E.

2. In this case, B and C send Signaling Route Set Test messages to D, requesting the status of the route to F. This occurs every 30 to 60 seconds until a transfer allowed message is received from D. (A transfer allowed message indicates that the destination has become available.)

3. Point D responds to the Signaling Route Set Test message with a transfer allowed, restricted or prohibited message as dictated by the current status.

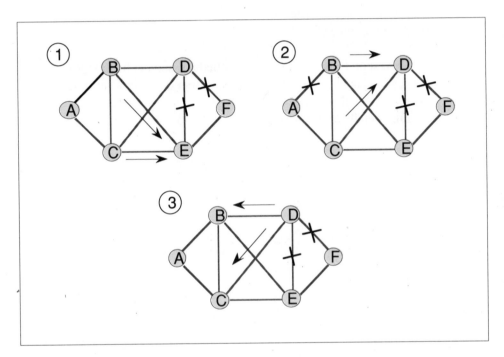

Figure 4.10 - Signaling Route Set Test

Transfer controlled: This procedure is used to notify adjacent signaling points of congestion at a signaling point or the congestion status of a signaling route. The action taken on receipt of a transfer controlled message is network-dependent. The procedures here vary depending on the implementation.

Signaling Route Set Congestion Test: This test is performed at a signaling point to determine the congestion status of a specific destination. A signaling point may query the signaling route status at other signaling points by the signaling route set-test procedure.

Congestion information is transmitted and requested by the transfer controlled message and signaling route set congestion test message, respectively.

The following example lists the steps carried out, and they are illustrated in Figure 4.11 - Signaling Route Set Congestion Test.

1.   B sends the congestion test message to D querying the congestion status of the route B-D-F.

2.   D may respond with a transfer controlled message indicating the congestion status level. If D does not respond within a set time interval, then B will resend the congestion test message to D. Specific details of this procedure are network-dependent.

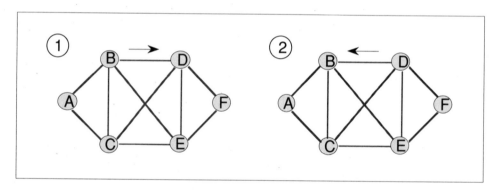

Figure 4.11 - Signaling Route Set Congestion Test

Signaling Traffic Management

The signaling traffic management functions are used to divert signaling traffic from a link or route to one or more different links or routes, or to temporarily slow down signaling traffic in the case of congestion at a signaling point.

The diversion of traffic in cases of unavailability, or availability, or restriction of signaling links and routes is typically made by means of the following basic procedures.

Signaling Point Restart: This procedure is used when a signaling point first becomes available, that is, when it establishes an active signaling link with another signaling point in the SS7 network.

When the first signaling link of a signaling link set becomes active, signaling traffic destined for the remote end of the link set is restarted.

As signaling links become active, transfer allowed, restricted, and prohibited messages are exchanged between the restarting and adjacent signaling points.

Management Inhibiting: This procedure makes a signaling link unavailable to user part traffic, but still allows maintenance and testing traffic to be carried out on the signaling link. This may be necessary where reliability problems are being experienced on the signaling link.

A request to inhibit a signaling link is not carried out if it means that an accessible destination would become inaccessible or if congestion was currently a problem in that part of the SS7 network.

Signaling Traffic Flow Control: This procedure restricts signaling traffic at its source when network failures or congestion occur. A number of conditions will trigger flow control action:

1. Signaling Route Set Unavailability: When the MTP determines that there is no signaling route available for a particular destination, the MTP informs the local user parts that signaling traffic destined for that destination cannot be transferred via the SS7 network. The local user parts take appropriate steps to stop placing signaling traffic for that destination on the network.

2. Signaling Route Set Availability: When a transfer allowed message informs an MTP that a signaling route is now available to a previously inaccessible signaling point, the MTP informs the local user parts that signaling traffic destined for that particular signaling point can be transferred via the SS7 network. The local user parts take appropriate steps to begin transmitting signaling traffic for that destination.

3. Signaling Route Set Congestion: When the MTP is informed that the signaling route to a particular destination is congested, the

MTP informs the local User Parts of the congestion. In networks with congestion priorities, the MTP informs the User Parts of the signaling point's congestion status.

4.   User Part Failure: If an MTP is unable to deliver a received message to a local User Part, the MTP sends a User Part unavailable message to the signaling point originating the message. The originating MTP, on receiving the User Part unavailable message, informs the affected local User Part of the failure at the remote User Part.

Forced Rerouting: This is the basic procedure to be used when a signaling route towards a given destination becomes unavailable. The objective of the forced rerouting procedure is to restore, as quickly as possible, the signaling capability toward a particular destination, in a way that minimizes the consequences of failure.

Forced rerouting is initiated at a signaling point when a transfer prohibited message, indicating a signaling route unavailability, is received. The following example lists the steps carried out, and they are illustrated in Figure 4.12 - Forced Rerouting.

1.   In the event that links D-F and D-E fail, D sends transfer prohibited messages to B and C, indicating that they must not route traffic destined for F via D.

2.   Transmission of signaling traffic towards F on the unavailable routes is immediately stopped, and such traffic is stored in forced rerouting buffers. An alternate route is determined according to the rules in Signaling Traffic Management. In this case, both B and C would select the links to E as the alternate route.

3.   As soon as Step 2 is completed, the signaling traffic destined for F is restarted on the alternate route link sets to E, starting with the contents of the forced rerouting buffer.

4.   B and C would send transfer allowed messages to D, indicating to D that messages destined for F could be routed via B and C.

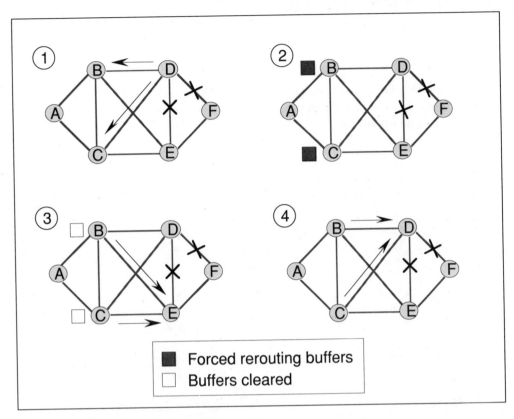

Figure 4.12 - Forced Rerouting

Controlled Rerouting: The controlled rerouting procedure is used to restore the optimal signaling route following certain types of changes in the status of signaling links and signaling routes.

Controlled Rerouting is activated on the receipt of a transfer restricted, or a transfer allowed message. The following example lists the steps carried out, and they are illustrated in Figure 4.13 - Controlled Rerouting.

1. In the previous example, we saw that with the failure of links D-F and D-E, B and C initiated the forced rerouting procedure to route traffic destined for F via E.

2. On the recovery of link D-F, D sends a transfer allowed message to B and C indicating that F can now be reached via D. If D is the primary route for signaling traffic originating at B or C and destined for F, then the controlled rerouting actions are initiated at B and C.

3.   Transmission of signaling traffic at B and C destined for F via E is stopped and placed into a controlled rerouting buffer. Transfer prohibited messages are sent from B and C to D, indicating to D that D is not to route signaling traffic destined for F via B or C. These transfer prohibited messages are used to ensure that routing loops do not occur.

4.   After a pre-set time interval (usually one second), the signaling traffic contained in the buffers at B and C is transmitted on signaling links to D.

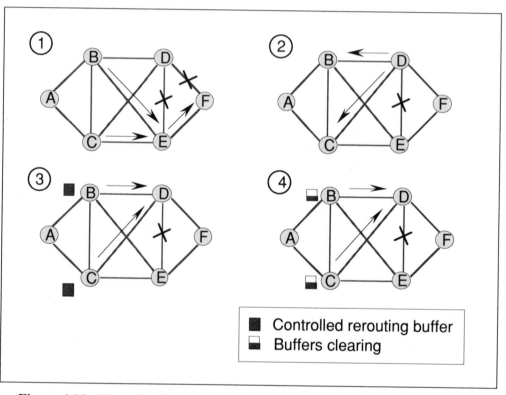

Figure 4.13 - Controlled Rerouting

Changeover: The changeover procedure is used to ensure that signaling traffic carried by an unavailable signaling link is diverted to the alternate signaling link(s) as quickly as possible while avoiding message loss, duplication or mis-sequencing. The changeover procedure includes buffer updating, which is performed before reopening the alternative signaling link(s) to the diverted traffic.

Buffer updating consists of identifying all those messages in the retransmission buffer of the unavailable signaling link which were transmitted but have not been received by the far end. The following example lists the steps carried out, and they are illustrated in Figure 4.14 - Changeover.

1. Suppose that a backhoe has damaged link A-B. Changeover messages are exchanged between A and B on an alternate route. These messages include information to allow A and B to update their buffers in order to retransmit those messages which were lost during the link failure.

2. Both signaling points respond to changeover messages with changeover acknowledgements.

3. Following receipt of the changeover acknowledgements, A and B now send traffic on the alternate route, beginning with the messages stored in the retransmission buffers.

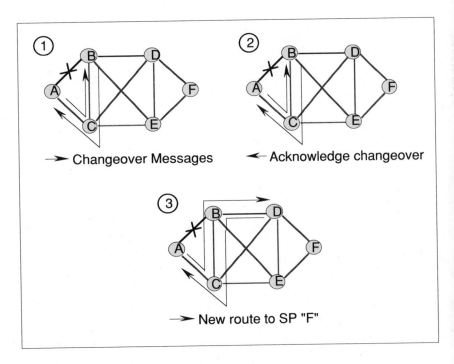

Figure 4.14 - Changeover

Changeback: The changeback procedure is used to ensure that signaling traffic is diverted from alternate signaling link(s) back to a link made available as quickly as possible while avoiding message loss.

The following example lists the steps carried out, and they are illustrated in Figure 4.15 - Changeback.

1. In the previous example, A and B initiated the changeover procedure when link A-B failed.

2. When the A-B signaling link becomes active again, a changeback message is issued by the signaling points at both ends of the link.

3. Both signaling points respond to changeback declarations with a changeback acknowledgement (ack).

4. Signaling points A and B immediately start sending traffic on the link made available.

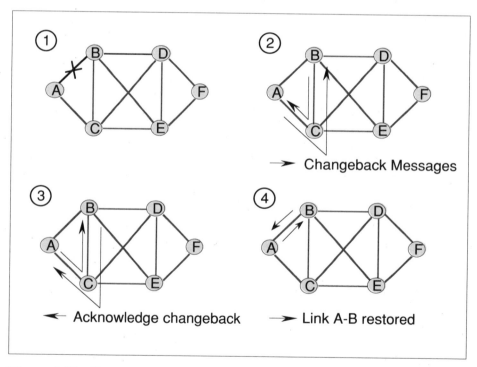

Figure 4.15 - Changeback

SIGNALING
5 CONNECTION
CONTROL PART

The Signaling Connection Control Part (SCCP) offers enhancements to the MTP Level 3 to provide connectionless and connection-oriented network services, as well as to address translation capabilities. The SCCP enhancements to the MTP services provide a network service which is equivalent to the OSI Network Layer 3, as illustrated in Figure 5.1 - SCCP in the SS7 Layers.

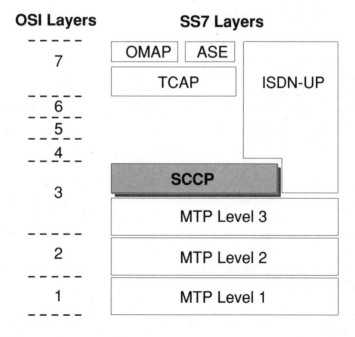

Figure 5.1 - SCCP in the SS7 Layers

SCCP ADDRESSING TRANSLATION

The routing capability of the MTP is limited to delivering messages to the correct signaling point based on the destination point code (DPC), and once there, forwarding the message to the correct MTP user within the signaling point based on the value of the service indicator contained within the signal information octet.

The SCCP provides an additional global title address translation function. A global title is an address, such as dialed digits for voice, data, ISDN or mobile networks, which cannot be routed on directly. The SCCP translates this number into a DPC and a sub-system number (SSN). The SSN identifies the SCCP user at a signaling point. Examples of SCCP users include SCCP management, ISDN-UP, and OMAP. The SSN is similar to the service indicator in the MTP routing but allows for 255 unique sub-systems to be defined at a signaling point, while the service indicator allows only 16 sub-systems to be defined.

CONNECTIONLESS SERVICE

The SCCP provides two classes of connectionless service. In both classes, the SCCP accepts signaling messages from SCCP users and transfers them across the signaling network as independent messages unrelated to any previously sent messages.

The basic connectionless service (Class 0 service) does not provide for segmenting/reassembly, flow control or in-sequence delivery.

The sequenced connectionless service (Class 1 service) is identical to Class 0 with one exception: it provides in-sequence delivery. If the SCCP user requests in-sequence delivery of a stream of messages, the SCCP sets the signal link selection (SLS) code to the same value for all messages in the message stream. While this doesn't guarantee in-sequence delivery, the MTP routing and error recovery procedures provide high probability that the stream of messages will be delivered in sequence.

The following example lists the steps carried out, and they are illustrated in Figure 5.2 - Connectionless Service.

1. When an SCCP user requests transfer of information using connectionless service, the SCCP function at the local SSP (A) creates a unit data message containing the information. The SSP_A transmits the unit data message to the remote SCCP contained in SSP (B). The information is then passed on to the SCCP user at that signaling point.

2. Additional information may be transmitted as required. There is no connection establishment or release associated with the connectionless procedures.

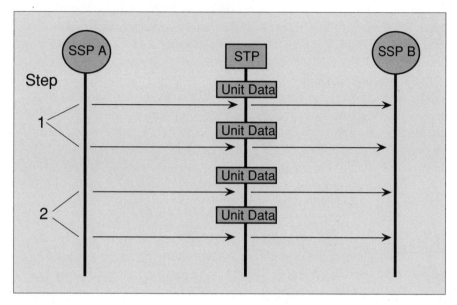

Figure 5.2 - Connectionless Service

CONNECTION-ORIENTED SERVICES

The SCCP provides two classes of connection-oriented services for the establishment of a temporary or permanent signaling connection to manage the transfer of messages between SCCP users. The signaling connection can be divided into three phases:

1. Connection Establishment: In this phase, a logical signaling connection is established between two SCCPs.

2. Data Transfer: Messages from SCCP users are exchanged across the signaling networks.

3. Connection Release: The signaling connection between two SCCPs is disconnected.

The basic connection-oriented (Class 2) service provides a bidirectional transfer of messages between two SCCP users. The SLS field is set to the same value in all the messages to ensure in-sequence delivery. This class also provides segmenting and reassembly of SCCP user messages. If an SCCP user delivers a message to the originating SCCP which exceeds 255 bytes, the originating SCCP segments the message into more than one SCCP data transfer message; it then transmits those data transfer messages to the destination SCCP which then reassembles the original message for delivery to the destination SCCP user.

The flow control connection-oriented (Class 3) service provides flow control where all messages are assigned sequence numbers and the SCCPs monitor the data transfer to ensure in-sequence delivery. In the event of mis-sequencing or message loss, the signaling connection is reset and the SCCP users are notified of the event.

The following example lists the steps carried out, and they are illustrated in Figure 5.3 - Connection-Oriented Service.

1. The connection-oriented procedures are characterized by the originating and destination SCCPs setting up a logical signaling connection. When an SCCP user requests a connection-oriented SCCP service, the originating SCCP function at SSP (A) creates a connection request (CR) message and transmits it to the destination SCCP located in SSP (B). The CR contains the relevant setup information including a source local reference (SLR) number (SLR = 14). The SLR is used by $SCCP_B$ in all subsequent messages to identify this signaling connection.

2. If $SCCP_B$ determines that the called party is a local user, and if resources are available, a connection confirmation (CC) is returned to $SCCP_A$. In the CC, $SCCP_B$ identifies the relevant signaling connection by referring to the original local reference (14) assigned by $SCCP_A$ in the CR. This is accomplished by setting the destination local reference (DLR) field to 14. $SCCP_B$ includes its own source local reference (SLR = 26). All subsequent messages related to this signaling connection include destination local reference numbers.

3. Data messages can now be exchanged between the two end-points of the connection.

4. When an SCCP user initiates a connection release, a released

message is transmitted to the remote SCCP. When the released message is received at the remote SCCP, an indication is sent to the SCCP user and a release complete message is sent back to the SCCP originating the release.

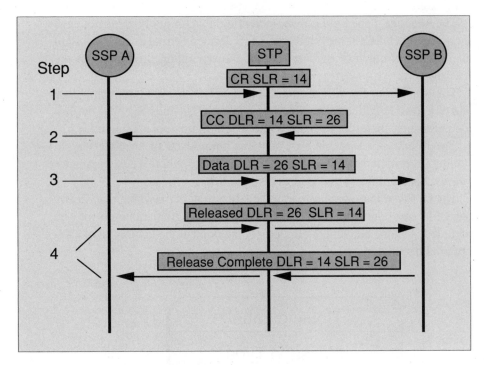

Figure 5.3 - Connection-Oriented Service

SCCP FORMATS

SCCP messages are carried on the signaling data link within the Signaling Information Field (SIF) of Message Signal Units (MSU), as illustrated in Figure 5.4 - Message Signal Unit.

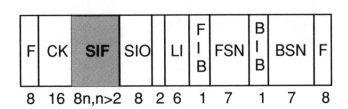

Figure 5.4 - Message Signal Unit

The SIF of each MSU containing an SCCP message consists of an integral number of octets composed of the parts illustrated in Figure 5.5 - Signaling Information Field.

Routing Labels contain the information necessary for the MTP to route a message.

The Message Type field is a one-octet field that uniquely identifies each message. Each SCCP message type has a defined format so that this field also identifies much of the structure of the remaining three parts of the message.

The Mandatory Fixed Part contains parameters that are both mandatory and of a fixed length for a specific message type. The message type defines the parameter, therefore the names and length indicators are excluded.

The Mandatory Variable Part contains parameters of variable length. Pointers are included in the message to indicate where each parameter begins. Each pointer is encoded as a single octet.

The Optional Part contains parameters that may or may not occur in any particular message type. Both fixed and variable length parameters may be included. A name and length indicator is included at the beginning of each optional parameter.

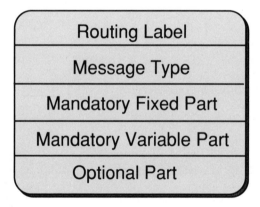

Figure 5.5 - Signaling Information Field

6 INTEGRATED SERVICES DIGITAL NETWORK USER PART

The basic function of the Integrated Services Digital Network User Part (ISDN-UP) is the control of circuit switched network connections between subscriber line exchange terminations. This includes basic voice and data services, and supplementary services.

Previous to developing the ISDN-UP, the standards groups developed the Telephone User Part (TUP) and the Data User Part (DUP). The ISDN-UP provides all the functions of both the TUP and the DUP, and is expected over time to replace their implementations.

The ISDN-UP provides some of the functionality of OSI Layers 3 through 7, as illustrated in Figure 6.1 - ISDN-UP in the SS7 Layers.

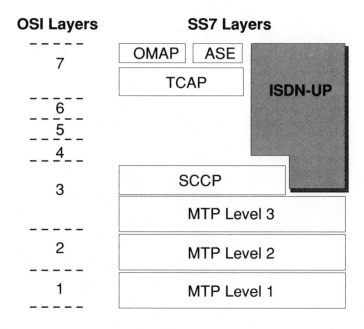

Figure 6.1 - ISDN-UP in the SS7 Layers

SERVICES

Basic Bearer Services

The basic ISDN-UP service is the control of 56 Kb/s or 64 Kb/s circuit-switched voice or data connections. This service, termed the basic bearer service is divided into three phases (in this way, it is similar to the SCCP connection-oriented service): call setup, connection, and call release.

Call Setup: The following example lists the steps carried out in call setup, and the steps are illustrated in Figure 6.2 - Call Setup.

1.   The ISDN-UP call setup procedures begin when the calling party initiates a call using the applicable access signaling. In this example, the calling party transmits an ISDN "setup" message.

2.   When the originating exchange has received the complete selection information from the calling party, and has determined that the call is to be routed to another exchange, selection of a suitable, free, inter-exchange circuit takes place. An Initial Address Message (IAM) containing information required to route the call to the destination exchange is sent by each exchange until the call reaches the destination exchange.

3.   The destination exchange, on receiving the IAM, notifies the called party of the incoming call using the appropriate access signal.

4.   The called party normally responds with an alerting indication which is passed backwards through the network as the Address Complete Message (ACM). When the originating exchange receives the ACM, an alerting message is passed to the calling party using the applicable access signal. At this point the caller hears the ringback tone.

5.   When the called party answers the call, a connect message is returned to the destination exchange, an answer message (ANM) is then passed backwards through the network. Call charges normally begin when the ANM is returned to the originating exchange.

6. When call setup is complete, a connect message is returned to the calling party.

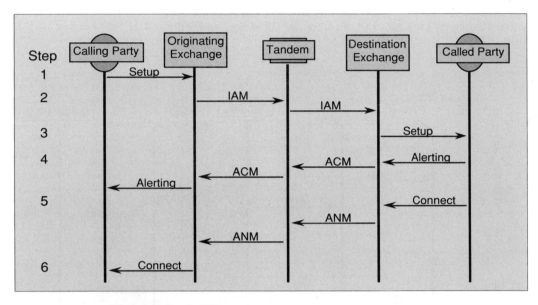

Figure 6.2 - Call Setup

Connection: End-to-end messages contain information which is required by the end-points of a circuit-switched connection. These end-points may be local exchanges or international gateway exchanges. End-to-end messages may be transmitted during the call setup or connection phase of the call. This phase is illustrated in Figure 6.3 - Connection Phase.

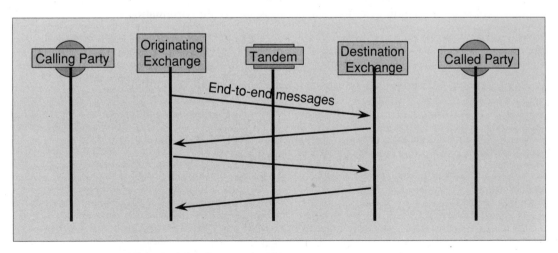

Figure 6.3 - Connection Phase

Call Release: The call release procedures are initiated when either end of a call sends a disconnect signal to the originating or terminating exchange using the applicable access signals. The following example lists the steps carried out in call release, and the steps are illustrated in Figure 6.4 - Call Release.

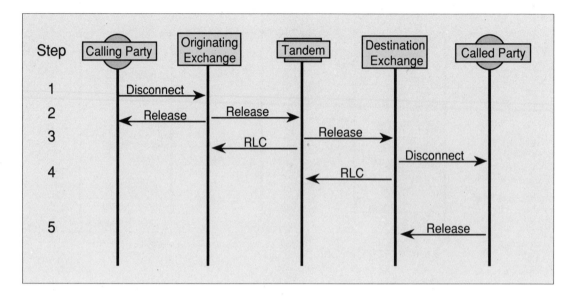

Figure 6.4 - Call Release

1.   In the example shown, the calling party has initiated the call release, by sending a disconnect message to the originating exchange.

2.   The originating exchange then sends a release message to the tandem exchange and returns a release message to the calling party.

3.   On receiving the release message, the tandem exchange returns a release complete (RLC) message back to the originating exchange and forwards the release message on to the destination exchange.

4.   When the destination exchange receives a release message, it forwards a disconnect message to the called party and returns an RLC message back to the tandem exchange.

5. On receipt of a disconnect message from the destination exchange, the called party returns a release message back to the destination exchange.

Supplementary Services

Redirection of Calls: This facility enables a user to have calls redirected to another predetermined number during periods when the facility is activated. This facility is illustrated here in Figure 6.5 - Redirection of Calls Facility.

1. In this example, user A activates call redirection to redirect calls to User B.

2. User C calls User A. The call is redirected to User B.

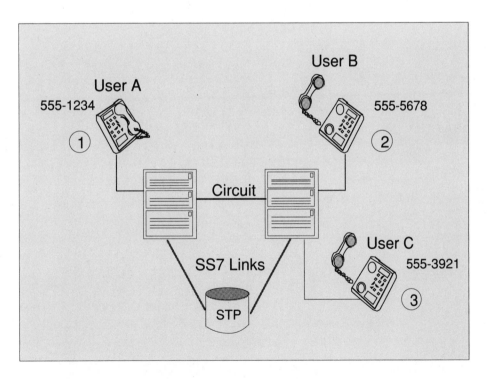

Figure 6.5 - Redirection of Calls Facility

3. The SS7 signaling network completes the call directly between User C and User B without tying up circuits at exchange A. This is the major advantage of call redirection over call forwarding. In the case of call forwarding, the call from User C would be routed to the switch serving User A and then tie up a second circuit when leaving switch A on its way to User B.

The Redirection of Calls Information Prohibited facility enables the user who has activated the redirection of calls facility to prevent the calling party from being informed that the call is redirected.

Malicious Caller Identification: This facility is a user-initiated request for the identification of the calling party and the called party. The following example lists the steps carried out in identifying a malicious caller, and the steps are illustrated in Figure 6.6 - Malicious Call ID.

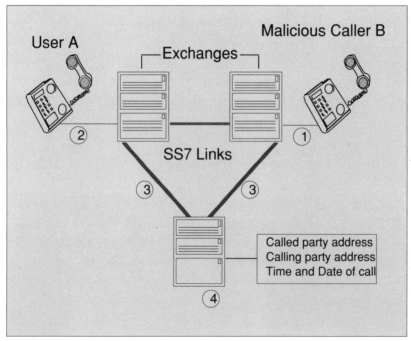

Figure 6.6 - Malicious Call ID

1. Malicious caller places a call from station B to User A.

2. User A initiates call trace once the call is recognized as a malicious call.

3. ISDN-UP sends the calling line ID of the malicious caller, called line ID of User A, and time and date of the call to the designated central location.

4. The information is stored for future investigation.

Calling Line Identification: With this service, the calling line ID is presented to the called party during call setup. This service is at the center of a "right-to-privacy" controversy where a number of groups are challenging the telephone companies' right to offer this service. In response to this controversy, some telephone companies are offering subscribers the option to restrict presentation of their telephone number to the called party.

Called Line Identification: This service provides the calling party with the the identity of the user to which the call has been connected. In the case where the call has been forwarded, the call forward reason is also displayed. The reason that the call was forwarded may be displayed for the caller's information; reasons for the call being forwarded may be that the line is busy, that there is no answer, or that all calls are being forwarded.

In this example illustrated in Figure 6.7 - Called Line ID, User A would have User B's number (555-5678) showing on his display. If the called party has the Address Presentation Restriction facility active, the called party address would not be presented to the calling party.

Closed User Group: A Closed User Group (CUG) facility enables users to form groups with different combinations of restrictions for access to or from users having one or more of these facilities. The following CUG facilities are standardized:

- closed user group with intercommunication only among members of the CUG;
- closed user group with outgoing access;
- closed user group with incoming access;
- incoming calls barred within the closed user group; and,
- outgoing calls barred within the closed user group.

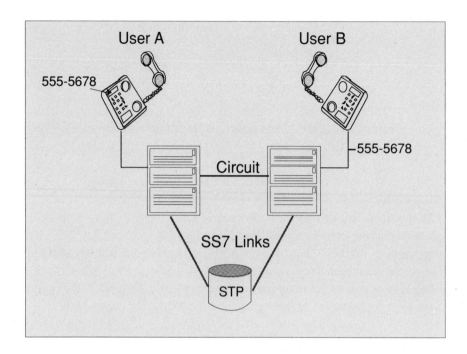

Figure 6.7 - Called Line ID

A user may belong to one or more CUGs. When this is the case, one CUG is nominated as the user's preferred CUG. Different combinations of CUG facilities may apply for different users belonging to the same CUG.

Completion of Calls to Busy Subscriber: The Completion of Calls to Busy Subscriber (CCBS) facility enables a calling party encountering the busy condition to complete the call automatically when the called party becomes free (without redialing). This facility is also known as network "ring again."

The calling party activates the user facility by making a request to the exchange to which he is connected. When the facility is activated, the status of the called party address is continually tested by its local exchange. When the address is free, the calling party is alerted. When the calling party answers, the call attempt is made once again.

ISDN-UP MESSAGE FORMATS

The ISDN-UP information is carried in the Signaling Information Field of the Message Signal Unit (MSU), as illustrated in Figure 6.8 - Message Signal Unit.

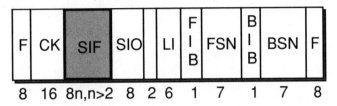

Figure 6.8 - Message Signal Unit

The Signaling Information Field consists of an integral number of octets that include the parts illustrated in Figure 6.9 - ISDN-UP Signaling Information Field.

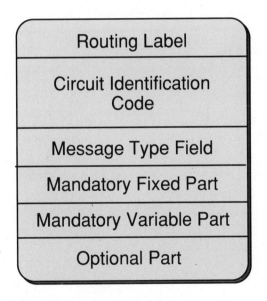

Figure 6.9 - ISDN-UP Signaling Information Field

Routing Label: This label contains the information necessary for the MTP to route the message.

Circuit Identification Code (CIC): This code identifies the circuit to be assigned to a call. The bits contained in the CIC are illustrated in Figure 6.10 - Circuit Identification Code.

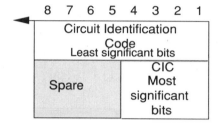

Figure 6.10 - Circuit Identification Code

For circuits that are derived from a 2048 Kb/s digital path, the Circuit Identification Code contains the five least significant bits, the binary coding of the time slot assigned to the speech circuit.

For circuits that are derived from an 8448 Kb/s digital path, the circuit identification code contains the seven least significant bits, the binary coding of the time slot assigned to the speech circuit.

In either case and when necessary, remaining bits are used to identify one among several systems interconnecting an originating and destination point. For circuits derived from other digital paths (for instance DS-1s in North America), similar CIC assignments are used.

Message Type Field: This field is a one-byte field identifying the message, for instance, Initial Address Message, Release Message.

Mandatory Fixed Part: Those parameters that are mandatory and of fixed length for a particular message type are contained in the Mandatory Fixed Part. The position, length and order of the parameters are uniquely defined by the message type, thus the names of the parameters and the length indicators are not included in the message.

Mandatory Variable Part: Mandatory parameters of variable length are included in the variable length mandatory part. Pointers are used to indicate the beginning of each parameter. Each pointer is encoded as a single octet. The name of each parameter and the order in which the pointers are set are implicit in the message type.

Optional Part: This part consists of parameters that may or may not occur in any particular message type. Both fixed length and variable length parameters may be included. Optional parameters may be transmitted in any order. Each optional parameter includes the parameter name (one octet) and the length indicator (one octet) followed by the parameter contents.

7 TRANSACTION CAPABILITIES APPLICATION PART

INTRODUCTION

The Transaction Capabilities Application Part (TCAP) is an SS7 application protocol which can be used by a variety of distributed applications. TCAP provides non-circuit related information transfer capabilities and generic services to applications, yet remains independent of the application. TCAP maps into the OSI Layer 7 as illustrated in Figure 7.1 - TCAP in the SS7 Layers.

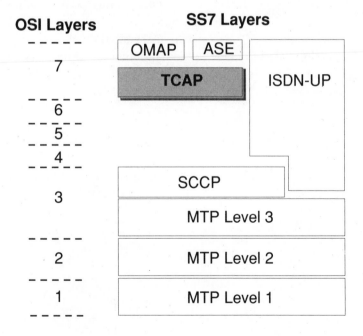

Figure 7.1 - TCAP in the SS7 Layers

Application process services (such as 800 service or network ring again) use TCAP to provide enhanced network services and operations, administration and maintenance (OAM) functions. An application process requiring services from TCAP is called a Transaction Capabilities User or TC-User. TCAP services may be used between:

- signaling points;
- signaling points and network service centers (such as databases or an OAM center); and,
- network service centers.

TCAP itself does not provide any services to telecommunications network users. Instead, it provides the capability for a large variety of distributed applications to invoke procedures at remote locations on the SS7 network. A common procedure is the query of a Service Control Point (SCP) database.

For example, 800 service uses the TCAP protocol to pass the dialed 800 number to an SCP database and request a translation to a routing number. The routing number is then returned to the signaling point to allow call routing.

TCAP services are based on a connectionless network service. Currently, no services are provided from the session, presentation or transport layers. TCAP interfaces directly with the SCCP, making use of the SCCP connectionless service to transfer information between TCAPs.

TCAP MESSAGE FORMATS

TCAP messages are contained within SCCP message signal units, as illustrated in Figure 7.2 - TCAP Message Signal Unit.

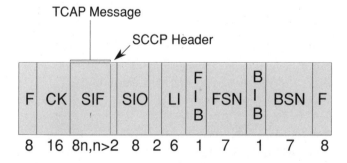

Figure 7.2 - TCAP Message Signal Unit

The TCAP message format consists of two parts, the transaction sub-layer and the component sub-layer, as illustrated in Figure 7.3 - TCAP Sub-Layers.

The component sub-layer is responsible for accepting components from TC-Users and delivering those components in order to the remote TC-User.

The transaction sub-layer is responsible for managing the exchange of messages containing components between two TCAP entities. This exchange of components, to perform an application, is called a dialogue.

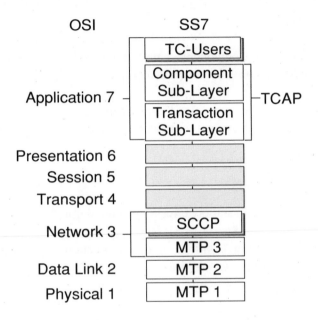

Figure 7.3 - TCAP Sub-layers

TCAP Component Layer Types

There are five types of components that may be present in the component portion of a TC message.

1. Invoke Component: This component requests that an operation be performed. It may be linked to another invocation previously sent by the other end.

2.   Return Result (not last) Component: When TCAP uses a connectionless network service, it may be necessary for the TCAP user to segment the result of an operation. In this case this component is used to convey each segment of the result except the last, which is conveyed in a Return Result (last) Component.

3.   Return Result (last) Component: This component reports successful completion of an operation. It may contain the last/only segment of a result.

4.   Return Error Component: This component reports that an operation has not been successfully completed.

5.   Reject Component: This component reports the receipt and rejection of an incorrect component, other than a Reject Component.

Transaction Sub-layer

There are five transaction sub-layer message types defined, as illustrated in Figure 7.4 - Transaction Sub-layer Messages. In order of transmission, the message types are: Begin, Continue, End, Unidirectional, and Abort. The begin, continue, and end messages are used in structured dialogues, while unidirectional are for unstructured dialogues, and abort is reserved for abnormal situations.

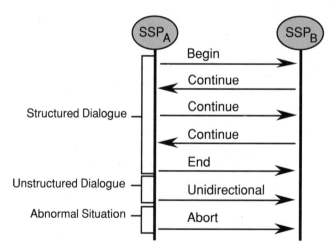

Figure 7.4 - Transaction Sub-layer Messages

The Begin message is used to establish a transaction with a remote peer transaction sub-layer. The Begin message may include one or more components.

The Continue message is used following a Begin message to transfer additional information related to the transaction. The Continue message contains one or more components.

The End message is used to terminate a transaction. All remaining untransmitted components are included in the End message.

Unidirectional messages are associated with unstructured dialogue; such a dialogue occurs when a TC-user needs to send one or more components to a remote TC-user and replies are not expected.

The Abort message is used to terminate a transaction following an abnormal situation detected by a transaction sub-layer or a request by the component sub-layer to abort the transaction.

Information Elements

The transaction and component sub-layers are structured in what are called information elements. An information element consists of a tag octet, a length octet, and a variable length contents field. The tag identifies the information element; the length specifies the number of octets in the information element, excluding the tag or length octets; and the contents contains the information that the element is intended to convey.

There are two structures that an information element may have: a primitive form, and a constructor form. Both are illustrated in Figure 7.5 - Information Element Forms. A primitive form is one whose contents field does not contain additional information elements. A constructor form is one where the contents field contains one or more information elements. These information elements may in turn themselves be constructor form.

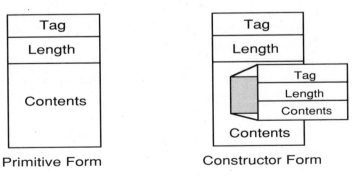

Figure 7.5 - Information Element Forms

TCAP APPLICATIONS

TCAP makes possible a wide variety of advanced network services based on information exchange between network components.

Network Ring Again

Network ring again can be invoked after a caller from one switch encounters a busy signal when calling someone on another switch. This feature allows the call to be re-established when the called party hangs up. The following example lists the steps carried out, and they are illustrated in Figure 7.6 - Network Ring Again.

1. User A places a call to User B. IAMs are sent to the destination exchange. Call Setup was described in detail in Chapter Six.

2. A Release (REL) message is returned to User A indicating that user B is busy. RLCs are returned to complete the release sequence.

3. User A may request Network Ring Again in which case a TCAP message requesting ring again is sent to exchange B.

4. User A's Ring Again request is acknowledged by exchange B.

5. Exchange B monitors the busy/idle status of User B's line. When User B hangs up, exchange B returns a TCAP station idle message back to exchange A.

6. Exchange A returns a TCAP end message to complete the TCAP dialog.

7. Exchange A alerts User A with a distinctive ring, and if User A picks up the phone, exchange A proceeds to re-establish the call to User B.

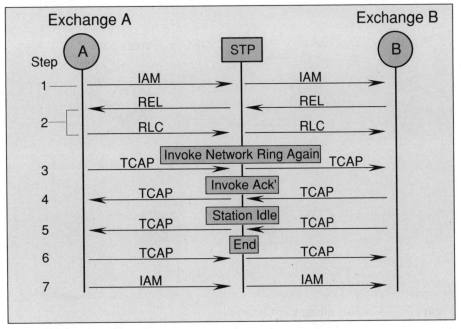

Figure 7.6 - Network Ring Again

Credit/Calling Card Verification

Credit/Calling Card Verification allows callers away from home or the office to place toll calls with the convenience of billing the call to a credit/calling card.

The following example lists the steps carried out, and they are illustrated in Figure 7.7 - Credit/Calling Card Example.

1. A caller dials a number with a request to bill the call to a credit/calling card.

2. Exchange A sends a request to an SCP containing the credit/calling card database to verify that the card is valid.

3. The SCP credit/calling card database verifies the card number.

4. A reply back to exchange A indicates whether or not the card number is valid.

5. If valid, Exchange A proceeds to place the call.

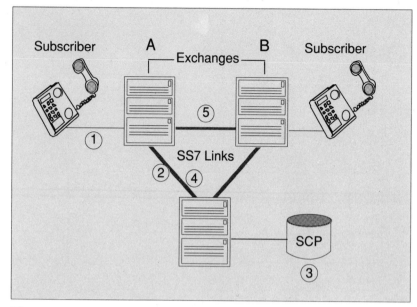

Figure 7.7 - Credit/Calling Card Example

800 Service

Enhanced 800 service allows telecommunications carriers to offer a number of flexible 800 services tailored to customer requirements.

Based on the time of day or the location of the caller, 800 calls to a specific 800 number could be routed to different locations. This arrangement would allow multi-location businesses or governments to tailor the services they provide to their customers.

The following example lists the steps carried out, and they are illustrated in Figure 7.8 - 800 Service.

1. A caller at exchange A dials an 800 number.

2. Exchange A sends a TCAP message to the 800 database requesting call routing information.

3. The 800 database located at an SCP determines the routing information for the call.

4.   The SCP sends the routing information back to exchange A.

5.   Exchange A then proceeds to set up the call.

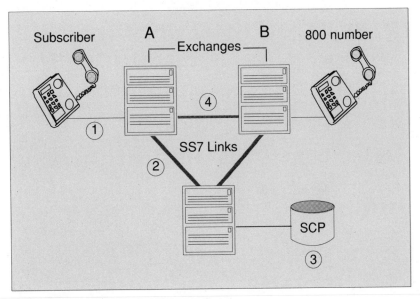

Figure 7.8 - 800 Service

OPERATIONS, MAINTENANCE AND ADMINISTRATION PART

8

The Operations, Maintenance and Administration Part (OMAP) provides procedures related to operations and maintenance functions. OMAP corresponds to the OSI model's Application Layer (Layer 7), as illustrated in Figure 8.1 - OMAP in the SS7 Protocol.

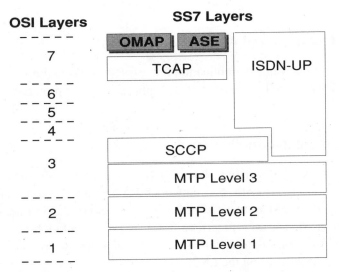

Figure 8.1 - OMAP in the SS7 Protocol

MANAGEMENT MODEL

The SS7 management model, in Figure 8.2 - SS7 Management Model, illustrates the relationship of the various management components.

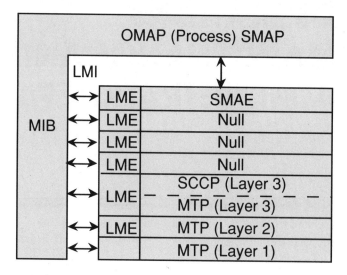

Figure 8.2 - SS7 Management Model

System Management Application Process

The System Management Application Process (SMAP) monitors, controls and coordinates resources through the Application Layer protocols.

Management Information Base

The Management Information Base (MIB) contains the performance and alarm data collected from the SS7 network by OMAP and the configuration data accessed by OMAP to assist in the management of the SS7 network.

The Layer Management Interface (LMI) is implementation-dependent and not subject to standardization. Data collected from the protocol layers is stored in the MIB using the LMI.

The Layer Management Entity (LME) deals with management functions of the corresponding SS7 layer.

System Management Application Entity

The System Management Application Entity (SMAE) is the aspect of SMAP involved with communications. SMAE contains one or more communications functions for an application. Each communications function is called an Application Service Element (ASE).

OMAP APPLICATION SERVICE ELEMENTS

Two OMAP application service elements (ASE) are defined: the Message Routing Verification Test, and the Circuit Validation Test. The OMAP ASEs use the services of TCAP to perform their functions.

Message Routing Verification Test

The Message Routing Verification Test (MRVT) is designed to verify that the routing tables in SS7 network nodes are consistent and that no routing loops or other anomalies are present. The following example lists the steps carried out, and they are illustrated in Figure 8.3 - Message Routing Validation Test.

1. At the initiating signaling point (A), a Message Route Verification Test (MRVT) message is sent on each signaling route which is contained in the MTP routing table for the test destination. The initiating signaling point then waits for Message Route Verification Acknowledgement (MRVA) and Message Route Verification Result (MRVR) messages to be returned.

2. At intermediate signaling points, if the test can be run and the initiating and test destination signaling points are known, an MRVT message is sent on each signaling route which is contained in the MTP routing table for the test destination.

3. Again, at intermediate signaling points, if the test can be run and the initiating and test destination signaling points are known, an MRVT message is sent on each signaling route which is contained in the MTP routing table for the test destination.

4. When the test destination receives an MRVT message, it checks the MRVT message to see if the initiator of the test is known and if a trace has been requested. If the initiator is known, this part of the overall test is considered successful. An MRVA message is returned to the signaling point(s) that sent the MRVT message(s) to the test destination. If a trace has been requested, the test destination also sends a MRVR message directly to the initiator of the test.

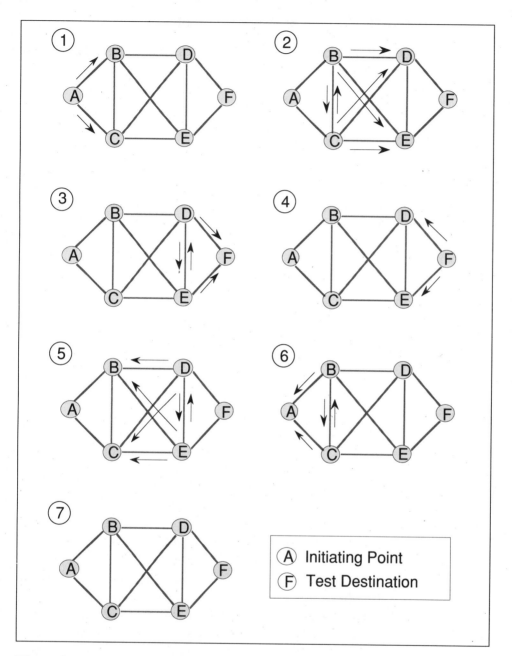

Figure 8.3 - Message Routing Validation Test

5. At the intermediate points, the MRVA messages received in step 4 are in response to previously sent MRVT messages from step 3. The information contained in the received MRVA messages is forwarded to the signaling point(s) that originally sent MRVT messages to the intermediate points. Under certain circumstances an MRVR message may be sent directly to the initiator of the test.

6. Again, at intermediate signaling points, the MRVA messages received are in response to the previously sent MRVT messages. The information contained in the received MRVA messages is forwarded to the signaling point(s) that originally sent MRVT messages to the intermediate point. Under certain circumstances an MRVR message may be sent directly to the initiator of the test.

7. The MRVT is now complete; test results are forwarded to the OMAP process requesting the test.

Circuit Validation Test

The Circuit Validation Test (CVT) is designed to ensure that two exchanges have consistent translation data for the circuits connecting the two exchanges. The following example lists the steps carried out, and they are illustrated in Figure 8.4 - Circuit Validation Test.

1. At the initiating end, an individual circuit to be tested is selected. Tests are then performed to ensure that data exists to:

    *   derive the physical circuit information for the circuit; and,
    *   derive the Circuit Identification Code (CIC) from the physical circuit information.

    If the near-end tests are successful, a CVT message is sent to the remote exchange.

2. The remote exchange receiving the CVT request message performs the following checks:

   a) Verifies that the CIC indicated in the CVT message is assigned;

   b) Ensures that translation data can derive a physical circuit from the CIC and routing label in the CVT message; and,

   c) Ensures that for the physical circuit, the CIC exists and can be derived from the physical circuit information.

3. If the far end tests are successful, a CVT message containing the derived CIC is returned to the initiating exchange.

4. At the initiating exchange, a comparison of the sent and received CICs are made. If they match, the test is considered successful.

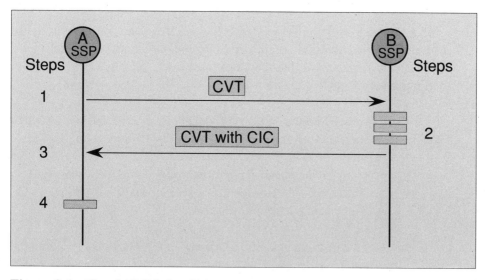

Figure 8.4 - Circuit Validation Test

9 APPLICATIONS

The ISDN-UP and TCAP sections of this book briefly covered some of the features and enhanced services made possible with SS7 signaling. In this section we take a look at some of these applications from a broader, end-user perspective. From this perspective, we stand to gain a better understanding of the benefits that SS7 signaling provides for both individuals and organizations.

The past decade witnessed a proliferation of features on single-site PBXs and Centrex platforms. However, over that period there was very little networking of these new services between switches. SS7 and adjunct services will enhance telecommunications capabilities by making features available on a network-wide basis.

VOICE MESSAGING INTEGRATION

Calling an associate in another city may result in the call being answered by a voice messaging system. You may be prompted by one of the voice recordings:

> "Press 1 to leave a message."
> "Press 2 to activate ring again."
> "Press 3 to speak to an operator."

Pressing "1" might store the calling line ID in the voice messaging system. When your busy associate checks her voice messaging system and hears your message, she may enter a one-digit code to transfer her call back to your number, simplifying the return call process. Pressing "2" would be a convenient way to activate Network Ring Again, while pressing "3" would provide access to an operator.

NETWORK AUTOMATIC CALL DISTRIBUTION

Network Automatic Call Distribution (NACD) is a natural fit for organizations such as airlines that have customer service centers located across the country over different time zones. NACD can improve customer service by pooling NACD agents at different locations into one large group. If one NACD location is very busy or out-of-service, incoming calls would be automatically routed to alternate sites. The same number of agents could now handle more calls, with a reduced average answering time. The airline gains sales, since more customers get through to agents and fewer customers hang up and call a competitor. The airline would also have more flexibility in selecting customer service locations and setting staff levels at each site.

A similar feature to NACD is Network Attendant Service where organizations can centralize operators at one site to minimize manpower.

VIRTUAL PRIVATE NETWORKING

Virtual Private Networking (VPN) is a service which provides multi-site organizations with the appearance of a private network with dedicated inter-switch circuits and a private dialing plan. However, in reality, the customer is part of a shared network with other VPN customers, or the customer is using the public network. VPNs provide users with a number of advantages of both private and public networks.

Organizations with multiple sites which are not large enough to support dedicated connections can use shared VPN facilities to connect the sites. However, the use of the shared VPN facilities still provides the advantages of a private network, such as access to long-haul facilities and services, private numbering plan, and feature transparency.

In the example illustrated in Figure 9.1 - Virtual Private Networking, users at sites A and C are linked to B with VPN facilities, permitting access to B's WATS and tie trunk facilities.

CUSTOM LOCAL AREA SIGNALING SERVICES

Custom Local Area Signaling Services (CLASS™) provides residential subscribers with many of the enhanced features previously available only with PBXs or Centrex service. The flexibility offered by these services can provide substantial benefits to busy users. From a distinctive ring for the in-laws, to the rejection of a particular telemarketing firm, there is a useful service for all.

CLASS is a registered trademark of Bell Communications Research.

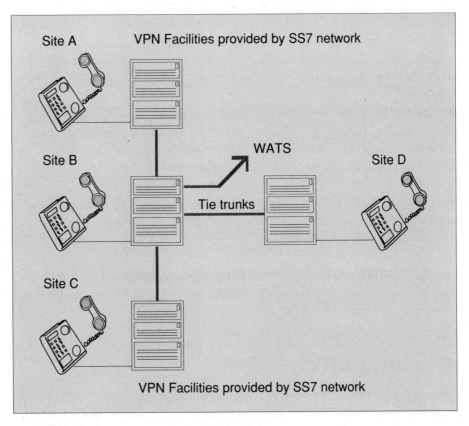

Figure 9.1 - Virtual Private Networking

Automatic Callback: Similar to a business's "ring again," automatic callback redials the last number called. When a connection is available, the originating party receives a distinctive ring, and the call is set up.

Automatic Recall: If you've ever made a mad dash to the phone, only to find the caller has hung up, automatic redial is the solution. Automatic redial, when activated, redials the last incoming call, without the user needing to know the number.

Calling Number ID: This feature shows the calling line ID on the phone display. This feature may be used to provide distinctive ringing, selective call forwarding, or selective call rejection.

Calling Number Display Blocking: Entering a special code blocks the calling party's number from being displayed on the called party's phone.

Customer Originated Trace: The subscriber can initiate a trace of the last incoming call. The caller's number is logged at the telephone company office for future reference.

Distinctive Ringing: A user-specified list of numbers is compared to incoming calls. Incoming calls with a matching number are indicated by a distinctive ring or call-waiting tone. Differentiating between a business partner and the teenager's friend is now simple and effective.

Selective Call Forwarding: The number of an incoming call is compared to a user-defined list of directory numbers. If the number matches, the call is forwarded to the special number.

Selective Call Rejection: Incoming call numbers are compared to a stored list. If an incoming call matches a number in the list, it is rejected (the subscriber's phone does not ring).

ENHANCED 800 SERVICE

Enhanced 800 service options provide organizations with considerable flexibility in tailoring 800 services to their unique requirements. In Chapter Seven - Transaction Capabilities Application Part, we discussed routing 800 calls based on day, time of call, and location of caller.

With the time-of-day and day-of-week routing options, organizations can set up full-time customer service centers backed up by part-time centers which only operate during peak business hours. All calls after hours, or on week-ends, would automatically be routed to the main business center.

Call routing based on location of caller provides additional flexibility. With this option not only could the call be routed based on the area code, but also on the NXX (the office code) as well. An organization could select an 800 number to be valid only in certain cities or regions within a state. When combined, 800 service with these enhanced features provide a powerful and flexible way to meet certain telecommunications requirements.

GLOSSARY

ACM	Acronym for address complete message.
Address complete message	An ISDN User Part trunk signaling message by which a call's destination SSP acknowledges an initial address message.
ANM	Acronym for answer message.
ANSI	Acronym for American National Standards Institute.
Answer message	An ISDN User Part trunk signaling message that indicates that the called party has answered the call.
Application layer	Layer 7 (the highest layer) of the OSI 7-Layer Reference Model.
ASE	Acronym for Application Services Elements.
Automatic callback	A CLASS™ feature that allows a subscriber to redial the last outgoing number automatically.
BELLCORE	Acronym for Bell Communications Research.
BIB	Acronym for backward indicator bits.
Bit	Individual data elements that make up bytes.
BSN	Acronym for backward sequence number.

Call forwarding	A POTS feature that allows a subscriber to forward calls from their number to another location.
Called line identification	In SS7, an address field that indicates the directory number of the called party.
Calling line identification	In SS7, an address field that indicates the directory number of the calling party.
CCITT	Acronym for Consultative Committee for International Telephone & Telegraph.
CCS7	Acronym for Common Channeling Signaling System no. 7.
CIC	Acronym for circuit identification code.
Circuit identification code	Used by the ISDN User Part to identify the message trunk associated with the call being signaled.
CLASS™	Acronym for Custom Local Access Signaling Services.
Common channel signaling	A signaling method whereby a channel conveys signaling information using labeled messages.
CUG	Acronym for closed user group.
Custom local access signaling services	The service mark of BELLCORE for a group of advanced subscriber services on the intelligent network.
CVT	Acronym for circuit validation test.
Data Link layer	Layer 2 of the OSI 7-Layer Reference Model.
Destination point code	A signaling point code that contains the network and member ID to which a message is addressed.

DPC   Acronym for destination point code.

DUP   Acronym for data user part.

FIB   Acronym for forward indicator bit.

FISU   Acronym for fill-in signal unit.

Global title   An address, such as dialed digits for voice, data, ISDN, mobile networks, which cannot be routed on directly.

GT   Acronym for global title.

IAM   Acronym for initial address message.

IAP   Acronym for initial alignment procedure.

Initial address message   In the ISDN User Part, the message used to initiate trunk signaling between two SSPs.

ISDN   Acronym for Integrated Services Digital Network.

ISDN User Part   The upper level (OSI Layer 7) protocol in SS7 that provides ISDN services.

Link set   Two or more signaling links that are connected to adjacent signaling points.

Line information data base   A database of subscriber information used to provide services.

Message Transfer Part   Layers 1 through 3 of the SS7 protocols, equivalent to the Physical, Data Link, and Network layers in the OSI stack.

MIB   Acronym for management information base.

MRVT   Acronym for management routing verification test.

MSU	Acronym for message signal unit.
MTP	Acronym for Message Transfer Part.
Multiplexer	Equipment that combines several signals into one higher-level signal.
NACD	Acronym for network automatic call distribution.
Network layer	Layer 3 of the OSI 7-Layer Reference Model.
OMAP	Acronym for Operations, Maintenance and Administration Part.
OPC	Acronym for originating point code.
Open Systems Interconnection	A standard for layered data communications to allow systems and terminals made by different suppliers to communicate together.
Originating point code	A signaling point code that identifies the signaling point where the message originated.
OSI	Acronym for Open Systems Interconnection.
Physical layer	Layer 1 (the lowest layer) of the OSI 7-Layer Reference Model.
Plain old telephone service	Voice telephone service without advanced features.
POTS	Acronym for plain old telephone service.
Presentation layer	Layer 6 of the OSI 7-Layer Reference Model.
SCCP	Acronym for Signaling Connection Control Part.
SCP	Acronym for service control point.
Service control point	An SS7 signaling point that provides a line information database for enhanced services and user information.

Service switching point	An SS7 signaling point located at a local or tandem switching office.
Session layer	Layer 5 of the OSI 7-Layer Reference Model.
SIF	Acronym for signaling information field.
Signal transfer point	An SS7 signaling point that acts as a packet switch.
Signaling	The process of controlling a network by exchanging messages between nodes.
Signaling Connection Control Part	One of SS7's upper level protocols, SCCP supports addressing and routing functions.
Signaling link	A physical and logical connection between two signaling points (in SS7, an STP, SSP or SCP).
SIO	Acronym for signal information octet.
SMAE	Acronym for system management application entity.
SMAP	Acronym for system management application process.
SS7	Acronym for Signaling System 7.
SSP	Acronym for service switching point.
STP	Acronym for signal transfer point.
SUERM	Acronym for signaling unit error rate monitor.
TCAP	Acronym for Transaction Capabilities Application Part.
Transaction Capabilities Application Part	An upper layer SS7 protocol that provides advanced services.

Transport layer	Layer 4 of the OSI 7-Layer Reference Model.
TUP	Acronym for telephone user part.
Unit data	An SCCP message that carries information for TCAP or ISDN User Part services.
VPN	Acronym for virtual private networking.

SUBJECT MATTER INDEX

ABOUT THE AUTHOR

Toni Beninger is Vice-President of Sanctuary Woods Inc., a leading interactive multimedia software company specializing in computer-based training for the telecommunications industry. Ms. Beninger earned her B.Sc. degree from the University of Ottawa, and subsequently completed her Education qualifications at Queen's University. She has had an extensive career in telecommunications documentation and training. Prior to joining Sanctuary Woods in 1989, she was the senior technical training manager at Bell-Northern Research Ltd.'s main laboratory in Ottawa.

As Vice-President of Sanctuary Woods, Ms. Beninger is responsible for the company's growing library of computer-based training courses, including SONET, Signaling System 7 (SS7), and Open Systems Interconnection (OSI). She is also responsible for the provision of the company's custom interactive multimedia services to the international telecommunications industry.

She may be reached at Sanctuary Woods Inc., Suite 501, 340 March Rd., Kanata, Ont., K2K 2E4 Canada; (613) 592-9784.